An Electric KAYAK

BUILD AN ENTRY LEVEL
ELECTRIC POWER BOAT - $500

CHARLES A. MATHYS

NETCAM PUBLISHING

Copyright © 2021 Charles A. Mathys

i

ISBN: 978-0-9843775-4-1

Cover design by: Charles Mathys and Cheryl Sivewright
Contribution Acknowledgement
Inside Graphics/Photos: by the author unless otherwise indicated

Dedication

*For my wife of 64 years who as a nurse spent her life
learning and teaching others. She inspired me to
continue learning about the many aspects of boat propulsion.
Over some 20 years, I have written 3 books on the subject
with her blessing and understanding.
I miss her generous nature, her companionship
and her advice on living a good life.*

Twenty years from now, you will be more disappointed
by the things you didn't do than those you did.
So throw off the bowlines. Sail away from safe harbor.
Catch the wind in your sails. Explore. Dream. Discover.

MARK TWAIN

Acknowledgements

I have many people to thank for their help working on this project and putting it in book form.

My grandson, Dan Robartes reviewed and edited an early manuscript of the book with much helpful advice.

To my son, Charles, who did a great job reviewing and editing the book and made many suggestions for its improvement.

Tim Knox of the Electric Boat Association who thought of the clever tittle "Up the creek without a paddle" for my electric kayak article. It gave me the impetus to turn the article into a do-it-yourself book.

My friend Bob McKelvie encouraged me to persevere with the project and gave me good advice for the composition of the book.

A big thank you to my granddaughter Hayley Robartes who helped me with the water testing and did the heavy lifting moving the kayaks around and back and forth to the local pond.

And, I am grateful to Cheryl Sivewright (outsource-design.com) who prepared the book for print, designed the cover and added a description of the book to my website: myelectricboats.com.

Contents

Introduction & Background

To make life a little easier for myself, I invested $200 for a "sit-upon" ten-foot kayak to replace my "sit-in" 10-foot kayak. At my very senior age, getting in and out of a sit-in kayak is quite a chore! As shown in the following image, I added a pair of sturdy handles and a home-made wood mast and boom. Since the boat is basically built like the boards that are so popular today, I thought that I could install a small sail and control the direction of the boat with the paddle. That did not work. I then added a weighted centerboard, but the board still did not sail well. I decided that building a sizable rudder would have to be the next step.

Sit-Upon 10-Foot Kayak with an Opti Sail Plan

While I was weighing the pros and cons of various sailing rigs, I spent a day at my favorite beach in Hingham, a town in Boston Harbor dating back to 1636. The beach is muddy at low tide, consequently, it is not a busy place. The town runs a sailing school nearby where they start the kids on Optimists (Optis) 8-foot skiffs. Their 35 square foot sail and 7.5-foot mast seemed about right for my new kayak.

The Opti was designed by an American, Clark Mills, in 1947. He built the largest pram that he could get out of two sheets of plywood. It went on to become one of the most popular sailing dinghies in the world. He donated the plans to the Optimist Club.

Getting back to my day at the beach, as I was walking to the sailing school to find out how the kids rigged their Opti, I noticed that there was not a bit of wind. I fully expected to find a group of kids with very long faces. On the contrary, they could not have been happier. Some of the kids were laying at the bottom of their boats, paddling with their hands, others were using the boat's scupper for water fights and some were just happy to jump in the water from the dock with their clothes and sneakers on. I could not help wondering why grown-ups can't enjoy a day on the water as much as the kids do. I think that the grown-ups could if boats were more reasonably priced (a new Sunfish costs $5000) and the boats were easy to transport, set up and launch.

During the water tests of the Opti sailing rig, I encountered a gusty day and capsized the kayak twice. Fortunately, I was not far from shore and the boat righted itself without taking on water. The incident reminded me that kayaking can be a dangerous sport. Before venturing out too far, it is wise to learn to right the boat and practice getting back on board, take a lesson at a sailing school if necessary. At that point it was late in the season, I put the sailing idea on hold and started my new winter project: **An Electric Kayak**.

CHAPTER 1.
Electric Propulsion
for Kayaks

Cost-effectiveness and Efficiency

Like 10-foot kayaks for $200 each, the two small Minn-Kota and Newport Vessels, 12-volt, trolling motors costing around $125 are hard to beat when it comes to value. They weigh less than 20 pounds and put out enough power to outrun a fast paddler. They are not efficient. The dynamometer tests that I performed on these motors show that they are about 65% efficient at full power and as we will see, far less efficient at lower speeds. With a battery as large as a car battery, the low efficiency is not a serious problem, however, for a kayak, a smaller battery and better efficiency are very desirable features. Fortunately, the miracle of Pulse Width Modulation (PWM) using a small Arduino Uno microprocessor (to be described later) provides very efficient propulsion.

Sit-in 10-Foot Kayak Ready for an Electric Trolling Motor

Twenty years ago, I did a lot of work on electric boats resulting in a couple of books. My most successful boat was "Sunny II," a converted 19-foot sailboat consisting of an O'Day Rhodes 19 hull with an efficient British Etek motor as described in my book "My Electric Boats". The boat was on a trailer so that the batteries could be charged before putting the boat in the water. Although it required much less work than removing the batteries to charge them, and was more reliable than using solar cells on the roof of the boat to charge them, dealing with large, heavy batteries was still the most unfulfilling part of my electric boating experience.

Much has been done over the last 20 years to improve the capacity and reduce the weight of batteries by using Lithium-Ion rather than lead. These improvements will continue because there is a growing market for electric cars and for cordless tools (Ryobi now has more than 175 different cordless tools). The Ryobi 18-volt, 4 Ampere hour (Ah) and 6 Ah battery packs are readily available everywhere.

We will see that by using PWM to generate the lowest trolling speed when the motor consumes only 35 watts, the continuous running time was measured at 2 hours from one 4 Ah battery pack. Of course, as the speed is increased requiring more power, the running time decreases. Considering the 2-hour charge time and the ease of plugging in a freshly charged battery pack in the control box to extent the cruise time, using a battery pack is an excellent solution for the battery problem.

The Proposed Project

The plan for the Electric Kayak project is to modify two 10-foot kayaks, the coral sit-upon kayak (without the sailing rig) and the red sit-in kayak. In addition to paddling them, it will be possible to propel them with 18 volt battery packs and slightly modified Minn-Kota or Newport Vessels trolling motors (we will recommend by-passing the rotary speed switch that comes with the motor but this is not a requirement). PWM techniques inherent in the Arduino Uno microprocessors are used to reduce the 18 volts of the battery packs to the 12 volts required by the motors and to control the speed of the motors much more efficiently. Ryobi 4 Ah and 6 Ah battery packs such as those used in cordless tools are used for the battery power.

By modifying 2 kayaks and driving them with 2 different motors we will be able to determine which combination of boats and motors provides the best performance for the intended use. My plan is to build the simplest and most DIY-friendly control box possible. Having tried several iterations of the control box, the final design will incorporate all the improvements discovered during the water tests. The mini transom and the steering changes should not be a problem for a competent DIY craftsman. If you have never worked with Arduino, it will be a challenging but worthwhile learning experience. The software has been written for you. It is simple and thoroughly checked out.

> ***Author's Note: Chapter 9 describes the use of off-the-shelf motor controllers which require no programming nor soldering of wires or components. Chapter 10 describes the use of more powerful 12 volt batteries for higher performance in speed and range.***

CHAPTER 2.
Project Overview

The Control Box and Other
Modifications to the Two Kayaks

To describe the project, I will use photos of the red sit-in kayak since I showed the coral sit-upon kayak as a sailing kayak. Photos of the main components, namely, the steering, the console and the mini-transom are shown below. I made the same electrical modifications to both kayaks and both motors. The boats can operate with either motor. With water test results from all four configurations, we'll be able to understand the advantages and disadvantages of each hull shape and of each motor.

Motor Mounted on a Mini Transom with Steering Lines

Mounting the Motor

Starting from the back of the kayak, the location of the mini transom from which we will hang the motor, is two inches from the end of the boat. In the past, I built a bracket for a small motor used on a canoe. The bracket was located near the seat so that there was no need to relocate the controls or to add steering. It worked but the handling (especially the turns) felt awkward, and I did not want to repeat those disappointing results.

In my distant past, I had seen large dories used by lobster fishermen fitted with a box inside the boat on which an outboard motor was mounted. The motor controls were used, and the steering was done with the outboard. All very convenient and easy to implement. Since I had already cut a hole for the centerboard in the coral kayak, I reasoned that by enlarging it, I could mount the trolling motor there within easy reach of the paddler. This idea did not work because the motor was mounted too close to the center of the boat for effective steering. On the other hand, the in-the-water tests did reveal the following good news: when the motor is mounted on the transom, both the controls and the steering work perfectly.

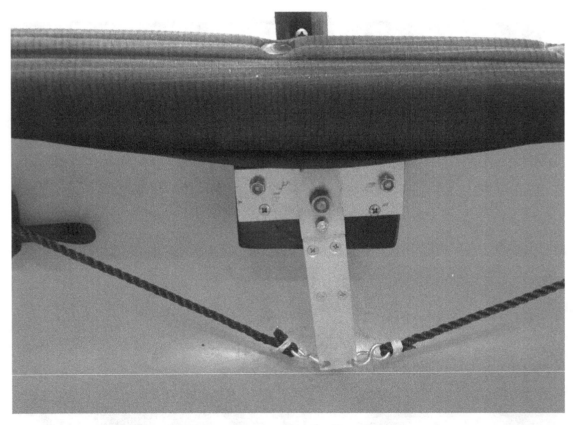

Pivot Arm Connected to the Steering Lines

The Steering

The steering is done with a lever on a pivot (shown in the preceding image) within easy reach of the paddler. It actuates a steering bar added to the motor as shown in the previous photo. A smooth ¼ inch line is attached to the end of the steering lever and runs out of sight on each side of the boat and emerges a couple of feet forward of the motor to connect to the steering bar. In front, it goes around a small pulley, the rest of the way it is guided through 6 eye bolts.

The steering can also be done using foot pedals. They can be purchased as a kit at Amazon for about $140. They were not tested for this book.

The Control Box

Much more will be said about the control box in Chapter 4 but in general, it must be located within easy reach of the operator. As a first line of defense for keeping it dry, it is mounted on a 3-inch-high pod. It houses a voltmeter, the Arduino Uno microprocessor and the Cytron driver. The two power wires for the motor (at least #14 gauge) go from the control box to the transom. These additions and modifications do not hamper the operation of the kayak with paddles.

Of course, the control box should be kept as dry as possible. An easy solution is to cover it with a plastic bag approximately 1 ½ gallons in size, with drawstrings at the bottom to secure it. A photo of the control box and the steering handle is shown below with a more complete description in Chapter 7.

Control Box on a Pod and the Steering Handle

Mini-Transom with Motor Power Terminals

The Mini-Transom

A transom is needed to mount the electric motor on the kayak. There is just enough room for a custom built mini-transom 7 to 10 inches wide (depending on the width of the kayak) to accommodate the motor clamps which require a minimum of 6 inches in width. To attach the transom to the boat, I used ¼ inch stainless steel threaded rods which do double duty by bringing the power to the motor from the control box to the surface of the transom.

Chapter 7 provides all the details needed to build and mount the steering equipment, and the mini-transom. We will get back to the construction and the wiring of the control box after we discuss the subject of *Pulse Width Modulation* which comes next.

CHAPTER 3.
Pulse Width Modulation

What It Is and Why We Need It

To take advantage of the inherent value of the Minn-Kota and Newport Vessel trolling motors, a 12-volt power source must be provided. But Lithium-Ion battery packs which have so many advantages over the lead acid batteries, generate 18 volts. Using Pulse Width Modulation (PWM), the 18-volt power is easily and efficiently reduced to 12 volts. The other equally important job for the PWM control is to provide 6 forward speeds and 3 reverse speeds. PWM control techniques will also make a dramatic improvement in the efficiency of these motors at low speeds.

Let's explore how the speed control of these motors normally works. Two resistors are connected to a complicated (and not very reliable) rotary switch to provide the 5 forward and 3 reverse speeds. These resistors are located in the front part of the motor, in the water, to dissipate the heat that they generate. For the slowest speed, the two resistors are connected in series and dissipate the most heat while providing the lowest battery voltage to the motor. Assuming that 4 volts go to the motor and 8 volts are wasted in heat (33% efficiency), the full speed 65% efficiency of the motor is now be further reduced to less than 25%. On the other hand, the PWM control operates at 98% efficiency. We can have confidence in this efficiency value because the Cytrons' ventilated MOSFET drivers which generate the PWM output power do not become hot when in use. This solution provides the best possible overall efficiency at all speeds.

The Arduino Uno and the Cytron Driver

The Cytron driver ($20) rated 20 amps at 30 volts is designed to run robot motors at various speeds in the forward and reverse direction. It is specifically designed to interpret PWM signals (0 to +5 volts) from devices such as the Arduino Uno ($15) to drive permanent magnet motors such as our trolling motors. In our project, using the Cytron and the Arduino Uno, we can reduce the battery voltage and provide 9 motor speeds in the forward and reverse directions very efficiently. This is a match made in heaven!

Pulse Width Modulation

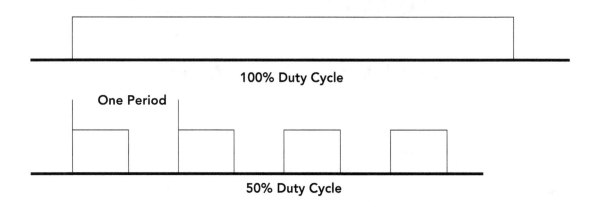

100% Duty Cycle

One Period

50% Duty Cycle

Switched vs PWM Control of the Motor

Let's explore how the PWM technique works. The diagram above shows the switched method for controlling the motor: on the top line (100% duty cycle) and the PWM method on the bottom line (50% duty cycle). The top line simply shows that a switch turns the motor power on and off. For example, if a 12-volt battery were connected in series with a switch, and 15 amps of current flowed when the switch is turned on, the motor would consume 180 watts of power, 100% of the time.

The bottom line represents the PWM control set at 50% duty cycle. The motor receives 15 amps for one half the time since the pulse width is set at 50%. In this example, it would consume 90 watts and generate one half the power. When an Arduino Uno is used, the duty cycle is determined by a PWM number from 0 to 255 which is interpreted by the software. For 50% duty cycle the number is 127 (50% of 255). If this motor were a trolling motor, it would push the boat at half speed. This reduced speed would be equivalent to the speed obtained when a stock motor's speed control handle is set to the middle of the five forward speeds. At that setting, with the series resistance inserted in the circuit, the motor would receive about 6 volts instead of 12 volts. As mentioned before, there is a big difference in efficiency between the two techniques: with the PWM, the drain from the battery is 90 watts whereas with the resistor method, the drain from the battery is about 150 watts — 60 watts are converted into heat by the resistors and wasted.

Pulse Width Modulation Tests

An oscilloscope was connected to the output of the Cytron driver which was driving one of the trolling motors. The images of the wave-shapes produced by the driver under PWM control are shown below. The first image shows the output at 50% duty cycle with the PWM set at 120. It correlates well with the first diagram that we discussed. The second image shows the output at the slowest speed (PWM = 60). These pulses are about half the width of the 50% pulses. At this slow speed setting, the motor consumes about 35 watts in the open water.

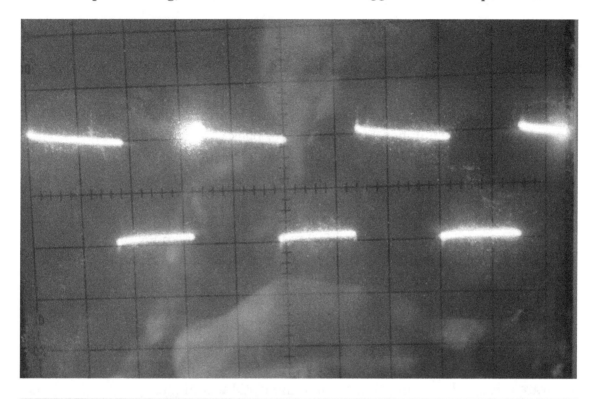

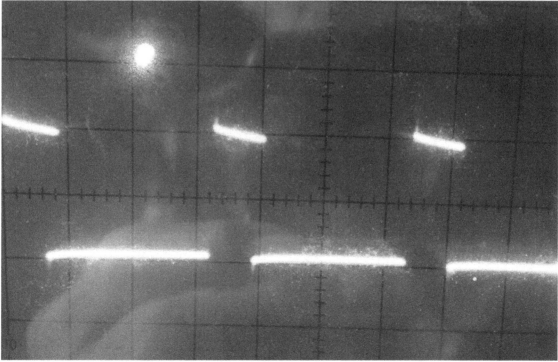

Oscilloscope Images

Driving a Motor under PWM Control

Wiring a motor with the Cytron driver and an Arduino Uno computer is a simple job. It takes just three control wires between the Uno and the Cytron, namely, Direction, PWM and Ground.

Programming the Uno for this task is equally easy. We will go into the details of writing the sketch (Arduino speak for "program") later, but at this point, let's consider the task of generating an appropriate signal for the PWM terminal of the Cytron driver.

Calculate the duty cycle for the Electric Kayak as follows. First, we calculate the full speed duty cycle number. Since our power pack voltage produces somewhat more than 18 volts, it takes 2/3 of 18 volts to run the motor at the full power equivalent of 12 volts. Converting this 2/3 value to a PWM number, a very good starting point is 160 (about 2/3 of 255). To obtain a range from full speed to trolling speed, we decrease this number in 20 point steps from 160 to 60 (PWM values of 160, 140, 120, 100, 80 and 60) in order to provide 6 forward speeds. This is a starting point, depending on the type of kayak, the water test results and the desired performance, the top speed and the intermediate speeds are completely adjustable by using any PWM values between 0 to 160.

That's all there is to defining the PWM signal. The oscilloscope images shown above produced the expected results. The PWM number used for the top photo was 120 and the number used for the bottom photo was 60. Until we describe the software this concludes our discussion of the PWM technique.

CHAPTER 4.

The Control Box

An Arduino Uno Microprocessor and a Cytron Driver Generate the Propulsion Power

The following image of the control box shows a front panel consisting of an on/off switch, an LED light, a speed control switch and an analog voltmeter (available at Amazon for about $12). On the same page, there is an image of another control box with a smaller digital voltmeter. A voltmeter which continuously keeps track of the condition of the battery is the main instrumentation necessary during a boat trip. With the smaller digital voltmeter, it is possible to reduce the width of the control box from 5 in. to 4 ½ in. That would entail rotating the Cytron board 90 degrees and raising the Uno so that it is mounted above the Cytron. On the right side of the box, a Ryobi power pack is shown, a 4 Ah battery is shown but a 6 Ah battery fits equally well. These battery packs display their charge level with a four light display at the bottom of the battery.

The control box houses the Arduino Uno microprocessor (attached to the front panel) and the Cytron driver attached to the bottom panel of the box. Not visible are two terminals (on the back panel) for the #14 gauge wires that are connected to the motor. The construction of the snap mount for the battery pack is described in the control box construction section below.

Construction of the Control Box

The custom-made control box is made (mostly) of ¼ inch plywood glued together. The front and back panels are screwed on to access the electronics. The diagrams below show the dimensions of the panels.

> ***Author's Note: An option to create the Control Box with a 3D printer is discussed at the end of Chapter 8.***

*Front View of the Control Box
with an Analog Voltmeter*

*Front View of the Control Box
with a Digital Voltmeter*

The basic design criterion is to make the box as narrow as possible to allow as much room as possible for the legs of the paddler. I have built four of these boxes and I can safely say that they can't be made much narrower than 5 inches without using a smaller voltmeter. In the last two designs, I mounted the components on a .040 aluminum panel. .040 in. thick aluminum is available at Amazon in 12 x 12 in. sheets for $12.

The Two Side Panels

Right Side Panel

Building the wooden box is straightforward except for the right side where the battery pack plugs into the box. To provide a solid backing for the battery pack, the thickness of the panel is doubled by gluing two ¼ inch plywood panels together. The sketch below shows the overall dimensions of this side panel and the detailed dimensions for the battery pack mounting. The photo which follows also shows the details of aluminum panels C and D with the indentures needed for the locking mechanism.

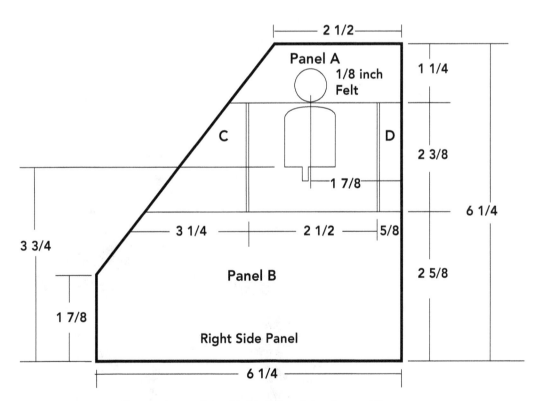

Dimensions of the Right Side of the Control Box

The two additional ¼ in. panels, namely panels A and B, are cut and glued on the right side panel, These panels provide a ¼ inch recess between them needed to support the battery properly. The bottom of the battery will rest on panel B but a ⅛-inch felt pad must be added to panel A (just above the cutout) to support the top part of the battery and keep it level.

The latching of the battery to the right panel is accomplished with two .040 inch aluminum plates (plates C and D made of the same metal as the front panel) and two plexiglass spacers .215 inches thick. The locking mechanism plates are exactly 2 ½ inches apart. The spacers underneath are 2 ⅝ inches apart to allow enough room for the snap locks to function. I happened to have plexiglass of the right thickness handy but if it is not available, the spacers can be made of ¼ inch plywood (which only measures .235 inch thick) by sanding about .020 from its surface.

The following image shows the latching arrangement. Note the numerous half circle cutouts that line up with the protrusions in the battery pack. They should be carefully measured and filed as shown with an ⅛ inch rattail file. The top of the hole is located just below the bottom of panel A and centered 1 ⅞ of an inch from the right edge. It has the same shape as the large protrusion from the battery pack where the contacts are located. Notice that the ⅛ inch groove at the bottom is not centered. A coping saw, sandpaper, files or a Dremel kit are all useful tools for cutting out the hole. Its location has to be scribed carefully and correctly. It is necessary to have a good fit so that the contacts will mate correctly and consistently with the battery contacts.

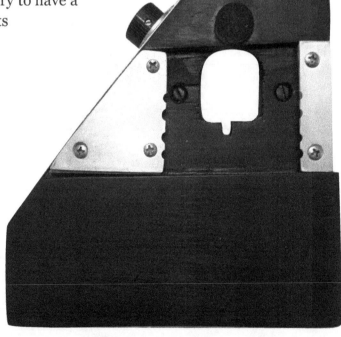

Right Side Panel

Battery Contacts

Making good battery contacts is not an easy task but a very important one. The photo above shows the heads of the 10-32 screws which hold the contacts firmly in place and serve as the connecting point for the #14 wire which goes to the + and – of the Cytron driver. The screws are located in the middle of the opening for the battery post about ½ in. from the edge of the hole.

A sketch of the contacts is shown in the diagram below. Note three important features: the contact area which is about one ¼ in. wide must press firmly in the middle of the battery contact area which measures 5/16 x ⅜ in.. The contacts must be bent as shown on the diagram so that they do not get hung up on the plastic surrounding the battery contact. Finally, it is important that once locked in place, the contacts don't move or twist out of position.

I tried unsuccessfully to find springy brass material to make the contacts. I did find some spring stainless steel that worked well. For another control box, I also broke up an old flashlight (with 2 D cells) that had ¼ in. wide conductors that could be shaped as a good, springy contact. The main problem was that there a minimal amount of material left where the hole for the 3-32 screw was drilled out. I put a dimple above and below the hole so that a flat washer could press the dimples into the wood and the contacts did stay firmly in place. But, a ⅜ in. or ½ in. wide contact is preferable.

The conductor from the flashlight was spring steel therefore, a wider ⅜ conductor made of spring steel about .020 thick, would work well where the screw is attached. The contact area must, of course, be ground down to ¼ in. as we see in the diagram. Stainless steel contacts are also a good choice even though their conductivity is much less than steel, the connections are so short that it does not matter and stainless steel has the advantage of staying corrosion free. So far, my best source for this material is a 3-inch automotive hose clamp with screw threads on one half its length. (The clamps sold at Home Depot do not work because they have screw threads cut in the metal from end to end).

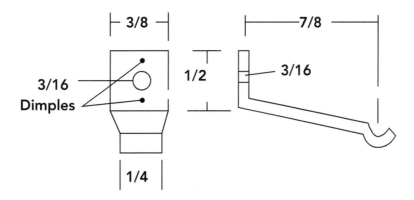

Battery Contacts (2 needed)

Left Side Panel

The left side panel is made of ¼ inch plywood and has the same overall shape as the right side panel but it is ½ in. shorter because the panel fits inside the top and bottom ¼ in. panels. The width is a ¼ in. less because it fits inside the back panel. To access the USB connector, a 1-inch hole is drilled at the location of the USB cable connector when the UNO is mounted in place. A one inch cap snaps over the hole to cover it when not in use.

One more detail. The left side panel has ⅜ x ½ inch reinforcing wooden strips glued to its four edges. This is done to increase the thickness of the ¼ inch panel where the front and back panels are attached with ¾ inch #6 screws. The control box mounting bars (¾ in. aluminum strips, 6 ¼ in. long), also screw into these reinforcing strips using # 6 flathead screws.

The Front Panel

The front panel is made of aluminum stock .040 inches thick. It is screwed on the wood strips ½ x ⅜ inches on the top, bottom and left side of the front opening. These strips are glued in place flush with the top, side, and bottom panels to provide a base for the four ¾ in., # 6 oval head screws that attach the panel to the control box. A sketch of the front panel with the analog voltmeter is shown below. It shows the suggested location of the components and the controls. The control box components include the following: a 30-volt analog voltmeter, the speed control rotary switch and the on/off switch with an LED indicator light to turn on the electronics.

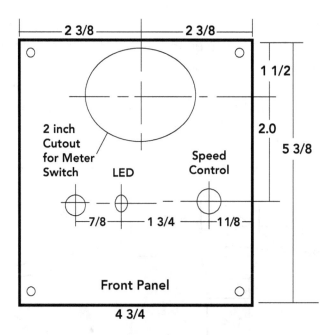

Front Panel with Analog Voltmeter and Other Components

The two places where the clearances are very tight are near the meter and near the rotary switch. If the meter is too close to the protrusion from the battery, it can be moved to the left $^1/_{16}$ to $\frac{1}{8}$ of an inch. It won't be quite centered but that won't be noticeable. By keeping the rotary switch as close as possible to the right side of the box, it should clear the Uno and the shield which, in turn, should be as close as possible to the left side of the box. A $^1/_{16}$ in. can be gained by letting the Uno's USB connector move into the 1 inch hole in the left panel which was drilled out and capped for access to the connector. When looking at the front of the panel, make sure that there is enough clearance between the speed indicator knob (1 $\frac{1}{2}$ in. in diameter) and the voltmeter case.

Below is a sketch of an alternate front panel with a smaller digital voltmeter ($6 at Amazon). By using the digital meter and rearranging the location of the other components, the control box can be made $\frac{1}{2}$ in. narrower with no tight spots. A bright green display for the voltmeter is best for operating in bright sunshine. A water-resistant version of this voltmeter is recommended.

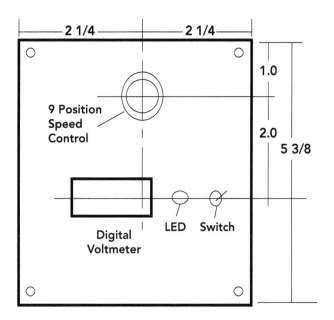

Front Panel with Digital Voltmeter and Other Components

The Other 4 Panels

The other 4 panels overlap the side panels in all cases. Consequently, they are ¼ to ½ inch longer or wider than the measurements of the side panels to which they attach. The dimension of the panels are provided but since they are simply rectangular pieces of ¼ in. plywood, a drawing is not provided. They must be hand sanded and the cracks need to be filled before painting. If the digital voltmeter is selected, the width of the panels is reduced by ½ inch.

Top Panel: 2 ½ inch x 5 inch wide

Bottom Front Panel: 1 ⅞ inch x 5 inch wide

Back Panel: 5 ¾ inch x 4 ¾ inch wide

Bottom Panel: 6 ¼ inch x 4 ¾ inch wide

The bottom panel fits between the side panels and on the inside of the front panel but it overlaps the back panel.

✎ *Major components used in this chapter are listed below.*

- 2 x 2 foot piece of ¼ inch plywood. Home Depot - $10
- Aluminum plate 12 x 12 in. .040 thick. Amazon - $12
- 12 position rotary switch with numbered dial (1 ½ inch in diameter) and On/Off switch with red LED. An electronics store such as All Electronics - $10
- 100 each ¼ watt resistors. Amazon - $6 or electronics store
- 30 volt DC Voltmeter (Hua, analog meter). Amazon -$12
- Citron Driver 20 Amps at 30 Volts. Amazon or Ebay - $20
- Arduino Uno Kit. Amazon or Ebay. $15
- Experimental shield for Uno. Amazon or Ebay - $ 3. Buy the one that is the size as the Uno. *Don't buy one that has screw terminals for the ports.*
- 4 Ah or combination of 4 and 6 Ah Ryobi battery packs. Two 4Ah packs. $100
- #14 braided wire with blue connectors, 10 feet
- #22 braided wire with red connectors, red, black, yellow, 5 feet each
- Miscellaneous stainless steel and brass screws.
- Front and back panel: 10 - #6, ¾ in., oval head.
- Mounting bar on the bottom of the control box: 4 flat head # 6, ¾ in. long and 4 oval head, #6, ¾ in. long. Brass screws for the battery contacts and the power terminals: 4 flathead, # 10/24 thread, 1 in. long with 8 brass nuts and 8 lock washers. Screws to attach Cytron to bottom panel: two ¾ in., flathead, #6. Screws to attach aluminum panels C and D: five ¾ in., flathead, #6.

CHAPTER 5.

Wiring of the Control Box

Diagrams and Instructions

Block Diagram

The block diagram of the control box pictured below shows the electronic connections to the speed control switch (10 wires) and the 3 wires between the two boards (PWM, Direction and Ground). The heavy lines from the Cytron driver show the two connections to the battery (be sure to observe the correct polarity!) and the two connections to the motor (the direction of rotation is determined by the polarity of these connections).

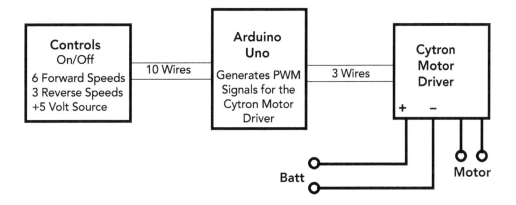

Block Diagram of the Control Box Wiring

The Power Diagram

The more detailed diagram below shows the power wiring, it includes the wiring from the battery pack to the Cytron driver board and to the motor. A second diagram (wiring of the electronics) shows how the controls are connected to the Arduino Uno microprocessor and to the Cytron driver. As wiring diagrams go, these two are very basic and easy to implement. You will need a soldering iron and solder to secure all wires and components such as resistors, switches and the LED. Do not exceed 25 or 30 watts and use the thinnest soldering wire available (.022 with rosin core). I like the lead based solder because it melts at a lower temperature.

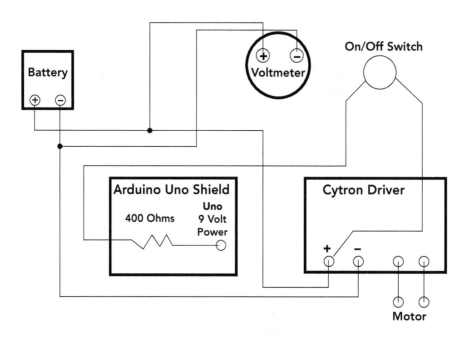

Power Wiring of the Control Box

Always use as little solder as possible! Clean the tip often and keep it tinned. Use a long thin tip. If you have a problem, clean the tip again and tin the parts to be joined.

If you don't have one, buy an inexpensive multi-meter such as "All Electronics" DVM-810 for $8. It will come in handy to check for continuity or the value of the components.

Starting at the + terminal of the battery pack or to the + of the Cytron driver, the power wiring shows the 3 connections to these terminals: a) to a 400 ohm, 2 watt resistor located on the Uno shield, b) to the + terminal of the Cytron and c) to the + 18 volt terminal post on the shield. The "a" connection's purpose is to provide power to the Uno. It simulates the use of a 9-volt battery. The resistor is used to reduce the 18-volt battery pack voltage to less than 12 volts in order to conform with the Uno specifications. From the resistor, a #22 red wire is soldered to the center terminal of the Uno's 9 volt battery connector.

The "b" connection is made to the + of the Cytron driver. As we have mentioned before the Cytron is not protected against improper polarity connections, it is therefore important to observe the polarity of the source. Note that there are no polarity marks for the motor. The reason is that the motor simply runs in the opposite direction when the wires are reversed. The wires should be connected so that the motor runs in the forward direction when the speed control switch is in one of the six forward positions.

The "c" connection goes from the Cytron + to the + 18 volt post on the shield. The 22-gauge wires (red if available) can be crimped along with the 14-gauge wire in the blue crimp connectors.

The - terminal of the battery goes to the - of the Cytron where a second wire (# 22, black if available) is inserted in the crimp and connected to the ground post on the Arduino shield.

The last two Cytron connections, go from the Cytron board to the motor using # 14 gauge wire. At the motor, the power connections can be made 2 ways. The easiest is to attach the 2 wires provided from the motor to the terminals on the transom. Then turn the control handle to position 5. That will connect the motor directly to the control box via the motor speed selector switch.

A better way is to modify the motor slightly. It involves by-passing the motor switch and the resistors. When the motor cover is removed, 4 wires can be seen going down the steering tube to the motor: 2 are white, they go to the resistors and can be ignored. The other two are heavier wires: red and black. These are the power connections to the motor. To by-pass the control switch, these 2 wires are cut and the ends going to the switch are taped up. The ends of the wires going to the motor are stripped and butt connected to two 18 inch long sections of # 14 power wire in order to reach the power terminals on the mini transom. That's all there is to the modification.

It avoids going through the motor control switch which now becomes inactive.

The Electronics Diagram

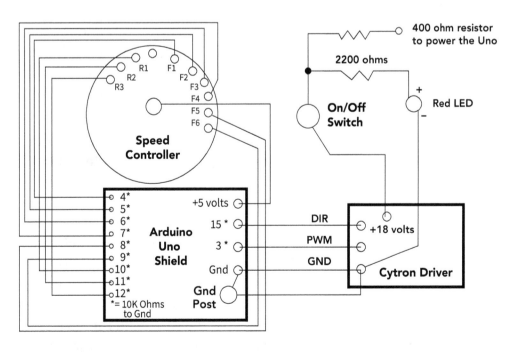

Electronic Wiring Diagram of the Control Box

We mention the word "shield" in the diagrams above. Many of the Arduino computers such as the Uno can accommodate a shield which plugs into the processor board with a pre-built circuit for special purposes such as connecting to the Internet. A prototype shield can also be purchased ($3) to implement custom circuits like these. It looks like a bare breadboard with a few connectors and usually a reset switch. We use one of these shields to conveniently mount the 10K resistors and connect wires.

The photo of the wiring side of the front panel shows the experimental shield with connectors on each side designed to plug into the Uno. They bring the Uno port signals to the shield board so that the signals can readily be accessed. Much of the board is filled with pins that are not connected to anything but can be turned into useful reference points such as the ground post and the +18 volt post. They are also used to anchor components to the board such as the 10 K ohm resistors. There are also numerous buses such as ground and +5 volts. We will use some of the 3 point buses that are not connected to anything. The pre-wired LEDs and switches mounted on the shield are not used for this application.

The electronic connections in the sketch above and in the photo below show all the interconnections between the speed controller, the on/off switch, the Uno shield and the Cytron driver.

The speed controller is a 12-position rotary switch. The center position is "Off". Turning the dial clockwise, selects speeds F1 to F6 one at a time. F1 is the slowest forward speed and F6 is full speed forward. When a speed is selected, the 5 volts from the center position of the rotary switch is connected to the selected pin. When the dial is rotated counterclockwise, reverse speeds R1 to R3 are selected one at a time. The speeds do not need to be selected sequentially and the motor can be put in reverse at any time. There is a slight delay (up to 1 second) between speeds. If your rotary switch has a stop, chose the "Off" position carefully to make sure that the stop does not interfere with the sequence of 6 forward or 3 reverse speeds.

The asterisks indicate that a 10,000 ohm, ¼ watt resistor is connected from ground to each of the 11 Uno ports that are used. This connection to ground is necessary so that the Uno port will not "float", an unstable condition which is neither a 1 nor a 0. All shields are not the same. The one to buy should be the same size as the Uno, have ground strips, +5 volt strips and unconnected three point buses. Connecting together the three wires from the rotary switch pin, the Uno port pin and one side of the 10K ohm resistor is not an easy task. In my first attempt to make these 3-way connections, I carefully slid the plastic housing from the shield connectors to expose the interior contacts, then I carefully soldered one end of the resistor to it. Then, I also soldered the wire from the speed controller to the resistor. The problem with this technique is that the soldering has to be done very quickly. If too much heat is applied to the exposed port contact, the plastic holding it upright melts and the pin falls sideways or against the next pin. In my second attempt, I used the jumper wires that came with the Uno computer. They can easily be plugged into the port pins. I am not a fan of plugging these tiny connectors into the port pins because the wires are so fine (# 30 gauge). They are hard to work with and tend to vibrate loose. A better solution is to use # 22 gauge *solid* wire which fits perfectly in the port connectors and can be substituted for the #30 gauge connectors.

The photo below shows a good way to install the solid wire from the port connectors to one end of the 3 point buses on the shield (there are 20 of them). The 11 white wires shown in the photo go to the nearest 3 point bus from the port connections (Uno ports 3* to 13*). The 9 input ports (4* to 12*) go to one end of the 3 point buses. At the other end of the 3 point bus a ¼ watt, 10 K ohm resistor is connected to a ground bus (there are two 12 point ground buses). The midpoint of the 3 point bus connects to the appropriate pin of the rotary switch using a yellow #22 gauge stranded wire. 9 rotary switch terminals are used for the ports. The common terminal of the rotary switch is connected to + 5 volts (on the shield) with a red wire.

Three more wires are needed for the three connections between the Cytron driver and ports 3*, 13* and ground (these ports are output ports). The jumper wires from the ports go to the nearest pin of two 3 point buses. The other end of the 3 point buses are connected to ¼ watt resistors connected to ground. The midpoint connection of the 3 point buses are connected to two #22 yellow wires twisted together with a black ground wire. They go to the Cytron screw connections for Direction, PWM and Ground. Tin the end of these #22 wires so that they don't unravel when the screws of the Cytron connector are tighten.

The last bits of electronic wiring are the connections to the on/off switch and the LED indicator. The same #22 stranded wire used above is fine for these connections. One side of the switch connects to the +18 volt post on the shield. The other side of the switch goes to two places: 1) nearby to a 2200 ohm resistor to limit the current to the LED (observe the LED polarity) and 2) on the shield, to the 400 ohm, 2 watt voltage dropping resistor which powers the Uno. The other leg of the LED is wired to the ground post on the shield.

The following image shows all the electronic connections. Though it may look complicated at first, when the wiring is done one section at a time, it is far easier than it looks. To review these sections:

- 11 white wires start at the Uno ports. Each one goes to a 3 point bus. At each bus one connection is a 10 K resistor which goes to ground. The other connection, in the form of a yellow wire, goes to one of two places. The 9 input ports go to the appropriate rotary switch terminals. The 2 output ports are twisted together with a black ground wire and go to the screw connectors of the Cytron

- The center terminal of the rotary switch goes to + 5 volts at the shield

- A black and a red wire come from the Cytron + and – and go to the + 18 volt post and the ground post on the shield

- A black and a red wire go from the shield posts to the + and – of the voltmeter

- A red wire goes from the + 18 volt post to one side of the on/off switch. The other terminal of the on/off switch goes to 2 places. 1) to the + side of the LED and 2) to the 400 ohm, 2 watt resistor on the shield. The other side of the 400 ohm resistor goes to the 9 volt connector of the Uno.

- A black wire goes from the – side of the LED to the ground post.

The Cytron and the Arduino Uno Boards

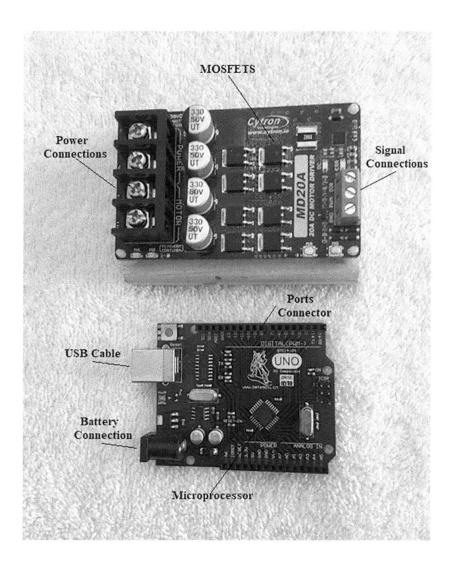

Referring to the image above, the shiny connector on the left of the Uno board is the USB connector. The black connector below is the 9-volt battery connector used to operate the Uno as a stand-alone unit (when it is not powered by the USB cable). We solder a wire from the 400 ohm, 2 watt resistor to the center point in back of this connector. On each side of the Uno we see the connectors into which the shield plugs in. The top row contains the digital ports from 0 to 13; some of these are PWM ports. The little black square in the middle of the Uno with many legs on each of its 4 sides is the microprocessor.

The Cytron driver above, has the power connections on the left and the electronic connections on the right (the green screw connectors). In the middle, are the Mosfet drivers which must be ventilated by lifting the board above its base.

The Uno is mounted on a 2 ½ x 2 ½ block of ¾ inch wood. It is attached to the front panel so as to clear the Cytron unit. It is located to the left as far as possible in order to minimize the interference with other components.

On top of the Uno, the experimental shield is held in place by the connectors on each side of the Uno. It is shown and described in the photo of the wiring side of the front panel. The large resistor mounted near the back of the shield is the 400 ohm, 2 watt resistor which provides the power to the Uno. In turn, the Uno generates the 5 volt power needed operate the microprocessor and ancillary equipment like the ports and the Cytron.

CHAPTER 6.
Software

The Arduino "Electric Kayak" Sketch
Controlling the Uno Microprocessor

The Arduino Uno was selected to control the Cytron driver because it is small yet powerful enough to perform the required tasks. The main requirement is to provide enough ports or channels for the 6 forward and 3 reverse speeds. It must also be able to generate a PWM signal to reduce the 18-volt battery pack voltage to 12 volts and to improve the efficiency of the motor at trolling speeds.

The program capable of doing this job (called a "sketch" in Arduino talk) is written on a PC and uploaded to the Uno board from a PC's USB connector. Before uploading the sketch, it is verified and compiled by the built-in compiler. The "Electric Kayak" sketch has been written for you so that it is merely a matter of typing a couple pages of code on your PC.

The Arduino IDE

When we refer to the Arduino IDE (Integrated Development Environment) we are talking about all the open source Arduino software that supports the development of thousands of different projects. To write and upload an Arduino sketch from your PC, you must first load this Arduino IDE comprehensive software program into your PC. Among its many features, are:

- A library of sketches
- An editor (similar to MS Word's) used to type in a new sketch using the "cut and paste" and "undo" features
- A Compiler which also verifies the sketch (and lets you know about it when you make a mistake!)
- The software needed to upload the sketch to a microprocessor board such as the UNO.

To download the latest version of the IDE, go to the Arduino website at https://www.arduino.cc/en/software. The IDE is an "open source" program which is completely free.

You will need to follow the installation instructions and select the USB drivers. You must also provide the name of your board (Uno) and the COM number of the PC port (usually 3). When you get to this point select a sketch such as "Blink" from the Arduino library (File/ Examples/ Basics/ Blink) to make sure that you can upload it to your Uno with the USB cable and run it.

You are now ready to copy the sketch "Electric Kayak" and upload it to the Uno. The control box will then be able to activate your trolling motor using the Cytron driver. Once the sketch is loaded into the Uno, it can be used over and over. Disconnecting the power to the control box or to the Uno board does not affect a program stored inside the Uno.

The Sketch "Electric Kayak"

If this is the first time that you have powered up your Uno board with a USB cable, you will immediately see a blinking LED on the board. This tells you that the board was shipped with a sketch already loaded in it. The blinking LED is the traditional "Hello World" symbol used to introduce a new program.

We described how to test the installation of the USB cable to the Uno board by uploading the sketch "Blink" to the board. A good short exercise at this point is to modify the delays (in "Blink") from 1000 (1 second) to 200 (1/5 of a second). After uploading the modified "Blink", you will see a very fast blinking LED. This exercise shows how easy it is to change parameters in a sketch.

The next job is to copy the sketch shown below into the program listings of the Arduino IDE. The procedure is as follows.

Open the IDE

When you open the IDE, under the heading "New", you will find a bare bones sketch which shows the two essential parts of any sketch: 1) the "setup" (which only runs once) and 2) the "loop" which runs over and over. When there are numbers (integers) in the sketch, they will need to be defined in the setup section. The "Electric Kayak" sketch replaces this bare bones sketch which is then deleted. Your first line of code starts with "int"

In the sketch "Electric Kayak", we have numbers (integers) corresponding to the ports. They are defined in the following example: "int KAYAK_PWM= 3;". This means that port 3 will generate the PWM outputs. In the setup section, we will show whether the port is used as an input or an output port. For example: "pinMode(3,OUTPUT);" In the loop section, the following example shows how we drive the motor according to a selected speed and direction: "analogWrite(KAYAK_PMW,100*val);"

The basic sketch which is to be deleted and replaced by the Electric Kayak sketch is shown below:

```
void setup() { // put your setup code here, to run once:
}void loop() { // put your main code here, to run repeatedly: }
```

Please note that the punctuation, the capital letters, the spaces, the underlining and the parentheses must all conform to the C++ rules. Any error will be picked up by the compiler and progress will stop until the error is corrected. It is very easy to make a punctuation error or omit a punctuation. Type the sketch with care! Use the editing functions "copy" and "paste" as much as possible to duplicate code that has passed the compiling test. Except for their PWM number, the six forward routines are identical and the 3 reverse routines are also identical.

The Electric Kayak sketch follows. A double slash (//) tells the compiler to ignore the rest of the explanatory information on that line. (It also comes in handy if you need to troubleshoot an error uncovered by the compiler).

```
// put your setup code here, to run once:
int KAYAK_PWM = 3;
int KAYAK_F1 = 4;
int KAYAK_F2 = 5;
int KAYAK_F3 = 6;
int KAYAK_F4 = 7;
int KAYAK_F5 = 8;
int KAYAK_F6 = 9;
int KAYAK_R1 = 10;
int KAYAK_R2 = 11;
int KAYAK_R3 = 12;
int KAYAK_DIR = 13;
int val = 0;
void setup ()    // () are the empty left and right parentheses
{
pinMode(3,OUTPUT);
pinMode(4,INPUT);
pinMode(5,INPUT);
pinMode(6,INPUT);
pinMode(7,INPUT);
pinMode(8,INPUT);
pinMode(9,INPUT);
pinMode(10,INPUT);
pinMode(11,INPUT);
pinMode(12,INPUT);
```

```
pinMode(13,OUTPUT);
}
void loop() {  // enter your main code (9 very similar speed routines) below, to run re-
peatedly
val=digitalRead(KAYAK_F1);
if (val == 1) {
digitalWrite(KAYAK_DIR,val);
analogWrite(KAYAK_PWM,60*val);
delay (1000);
digitalWrite(KAYAK_DIR,LOW);
digitalWrite(KAYAK_PWM,LOW);}

val=digitalRead(KAYAK_F2);
if (val == 1) {
digitalWrite(KAYAK_DIR,val);
analogWrite(KAYAK_PWM,80*val);
delay (1000);
digitalWrite(KAYAK_DIR,LOW);
digitalWrite(KAYAK_PWM,LOW);}

val=digitalRead(KAYAK_F3);
if (val == 1) {
digitalWrite(KAYAK_DIR,val);
analogWrite(KAYAK_PWM,100*val);
delay (1000);
digitalWrite(KAYAK_DIR,LOW);
digitalWrite(KAYAK_PWM,LOW);}
val=digitalRead(KAYAK_F4);
if (val == 1) {
digitalWrite(KAYAK_DIR,val);
analogWrite(KAYAK_PWM,120*val);
delay (1000);
digitalWrite(KAYAK_DIR,LOW);
digitalWrite(KAYAK_PWM,LOW);}

val=digitalRead(KAYAK_F5);
if (val == 1) {
digitalWrite(KAYAK_DIR,val);
analogWrite(KAYAK_PWM,140*val);
delay (1000);
digitalWrite(KAYAK_DIR,LOW);
```

```
digitalWrite(KAYAK_PWM,LOW);}

val=digitalRead(KAYAK_F6);
if (val == 1) {
digitalWrite(KAYAK_DIR,val);
analogWrite(KAYAK_PWM,160*val);
delay (1000);
digitalWrite(KAYAK_DIR,LOW);
digitalWrite(KAYAK_PWM,LOW);}

val=digitalRead(KAYAK_R1);
if (val == 1) {
digitalWrite(KAYAK_DIR,LOW);
analogWrite(KAYAK_PWM,60*val);
delay (1000);
digitalWrite(KAYAK_DIR,LOW);
digitalWrite(KAYAK_PWM,LOW);}

val=digitalRead(KAYAK_R2);
if (val == 1) {
digitalWrite(KAYAK_DIR,LOW);
analogWrite(KAYAK_PWM,80*val);
delay (1000);
digitalWrite(KAYAK_DIR,LOW);
digitalWrite(KAYAK_PWM,LOW);}

val=digitalRead(KAYAK_R3);
if (val == 1) {
digitalWrite(KAYAK_DIR,LOW);
analogWrite(KAYAK_PWM,120*val);
delay (1000);
digitalWrite(KAYAK_DIR,LOW);
digitalWrite(KAYAK_PWM,LOW);}
//delay (2000);
}
```

After typing in the sketch and checking that it compiles, save it with "save as" (under "File") with a file name of your choosing. To retrieve the sketch, the easiest way is by using the up arrow located at the top of all sketches. I verified the spelling of the sketch as it is spelled out above by having a third-party type in the sketch and compile it successfully.

How the "Electric Kayak" Code Works

The setup code shows that we have 2 output ports (ports 3 and 13). Port 3 is a PWM port, it is used to output the PWM signal to the Cytron driver. Port 13 is used to output the motor direction signal to the Cytron driver. Six speeds operate forward while 3 speeds run the motor in reverse. Ports 4 through 9 are the forward speeds while ports 10, 11 and 12 are the reverse speeds. "val" has a binary value of either one or zero. It becomes a 1 when that particular speed routine is selected by the rotary speed control switch.

The loop section is composed of 9 similar speed routines for the 6 forward and 3 reverse speeds. Each speed has its own PWM number and direction of rotation. For example, when the switch is turned one position clockwise (from the center/off position) the motor will operate at its slowest speed (PWM number 60) in the forward direction (F1). The next clockwise position will cause the second routine to become active. It will increase the speed to PWM 80 in the forward direction. The next 4 positions will continue to increase the speed in the forward direction. The last switch position runs the motor at its maximum speed with a PWM of 160. To activate the 3 reverse routines, the switch is returned to center/off and turned counterclockwise. Each step will increase the speed in reverse in the same manner as the speed was increased in the forward direction. The reverse routines are numbered 7, 8 and 9.

Looking at the 7 lines of code in each routine, we first determine which routine is active (if any). The first speed routine is speed F1 (Forward 1). The first line of code determines the value of "val". It can be a 1 or a 0 and it will only be a 1 if its switch position is selected. Center/off will return a 0 and so will any of the other speeds since only one speed can be selected at any one time with the rotary switch.

If val is a 0, we go on to the next speed routine. But if it is a one, meaning that the speed switch has a +5 volt on this contact, we will activate the next 5 lines of code following the open curly brace ("{") until we reach a close curly brace ("}").

The first of these lines of code "digitalWrite(KAYAK_DIR,val) means that we will activate port 13 with +5 volt to make the motor turn in the forward direction.

The second line of code "analogWrite(KAYAK_PWM, 60* val) means that we will activate port 3 (a PWM port) with a signal which causes a power pulse, 60/255 in duration, to be generated by the Cytron driver.

The third line simply means that we pause for 1 second (we let the motor run) under these operating conditions.

After the 1 second delay, we move to the fourth line of code, "digitalWrite(KAYAK_DIR,LOW)" it turns off port 13 and the fifth line of code, "digitalWrite(KAYAK_PWM,LOW)" which turns off port 3.

We now move to the next speed routine in the loop. The most likely situation is that the speed switch was not moved. After having checked the other 8 possible speeds, we find val = 1 again in the first speed routine. That situation will again cause the motor to run for one second before checking to see if the switch position has changed. This could go on for a very long time. There is no noticeable loss of power in the motor during this search. The Uno is so fast that going around the loop once (to check the other 8 speed routines) takes less than 50 microseconds, a time duration too short for the motor to notice.

Eventually, the speed will be changed or the speed switch will be returned to the center/off position. In that case, val will not be "1" in any of the speed routines and the motor will shut down. At this point, the power to the electronics can be turned off and the battery pack can be removed from the control box.

Summary of the Software Section

In this very interesting Chapter, we saw how the brains (the Uno computer) controls the battery pack muscles using the Cytron driver to run the trolling motor at various speeds in the forward and reverse directions. The Electric Kayak sketch is the program that controls the operation of the Uno's brains. We first setup the required 11 input and output ports. Then, using 9 similar "speed routines" we energized the trolling motor at the speed selected by the control switch. Simultaneously, the battery pack voltage, is reduced from 18 volts to 12 volts to conform to the operating voltage of the trolling motor.

CHAPTER 7.

The Main Components

Building and Installing the Steering, Transom and Control Box

My Electric Kayak is a stock 10-foot kayak with three major additions: a mini-transom on which to mount an electric trolling motor, a steering mechanism and a control box which operates the motor at various speeds in both directions and houses the battery pack. In this Chapter, I will describe how these additions are installed in the kayak.

The Control Box

The images shown in Chapter 2 and Chapter 4 are the latest versions of the control box with a single analog voltmeter. On the right side a Ryobi 4 or 6 Ampere-hour battery pack can be plugged in. On the left side, a 1-inch hole is provided so that a USB cable can be connected to the control box from a PC in order to modify or reload the program. In the back cover of the box there are 2 terminals for the two power wires to the motor.

In the red kayak, the control box is mounted on a 3-inch tall pod to avoid it getting wet from the water usually found in the bottom of a kayak. I made the pod out of 2 pieces of pressure treated 2 x 4 glued together. It measures approximately 3 x 5 inches. The pod's bottom is contoured like the bottom of the kayak. It is bolted to the bottom about 12 inches forward of the front of the seat with two ¼ inch flat head, stainless steel screws, 1 ¼ inches long. A ¾ inch board 3 ½ x 6 ¼ inch (ripped from a 2 x 4) is mounted across the pod with two 1 ¼ inch flat head, #8 stainless steel screws. The control box is then attached to the board with four #6 stainless steel, oval head screws ¾ inch long through the four ⅝ in. tabs which overhang the bottom of the control box. The tabs are the ends of two 6 ¼ inch aluminum bars ¾ in. wide which are attached to the bottom of the control box. (The first photo of Chapter 4 shows one of these tabs.)

For a sit-upon kayak the pod is not required, however, the 3½ x 6 ¼ inch board is still needed to raise the control box slightly. It is attached to the kayak with molly bolts. The control box is fastened to the 6 ¼ inch board as with the sit-in kayak.

The Steering Mechanism

The steering mechanism shown in the photo in Chapter 2 is simply an arm on a pivot mounted near the middle of the kayak (see the sketch below). Attached to the bottom of the arm is a rope that goes around the two sides of the boat and emerges about 2 feet from the trolling motor's steering bar. Pulling on the bar with the steering handle pulls on the right side of the motor's steering bar. It causes the boat to turn in the starboard direction.

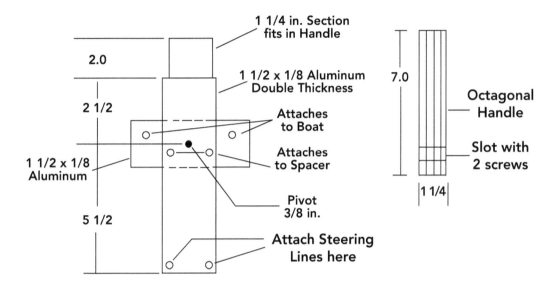

Dimensions of the Steering Mechanism

The construction of the steering mechanism is straightforward. The arm, is cut to size from two aluminum bars 1 ½ inches wide and ⅛ in. thick. The top 2 inches of these two bars are trimmed to 1 ¼ inches wide to match the diameter of the handle. The screws holding the handle on the bars will secure the bars together. They are also used to keep the bars aligned while the pivot hole is drilled. The length of the sections above and below the pivot point are 4.5 and 5.5 inches. Below the pivot the two bars are bolted together with 2 flat-head, #6 stainless steel, ½ inch long screws with lock-nuts. If desired, one of the two bars can be made 2 inches shorter than the other to make it easier to bent it slightly in order to obtain the maximum possible moment. Depending on the boat, it might be possible to increase the 5 ½ in. dimension somewhat in order to provide sharper turns for the kayak.

The pivot holes are ⅜-inch holes drilled 5 ½ inches from the bottom of the steering bar and in the center of a crossbar 1 ½ x 5 inches long. It accommodates a 1 ½ in. ⅜-inch bolt that has no threads for a distance of ⅜ inch from the head in order to provide a smooth rotating surface. The threaded part of the bolt is then cut flush with the end of the lock-nut. A washer (thin brass if possible) is inserted between the two aluminum plates for a smooth operation. A ⅝ in. hole ¼ in. deep is drilled in the spacer (described below) to accommodate the head of the pivot bolt. The aluminum crossbar is attached (with 2 # 8 stainless screws 1 in. long) to a wood spacer block which is also 5 inches wide and approximately 3 inches high and 2 inches thick.

The purpose of the 5 x 3 spacer block is to locate and attach the steering mechanism to the boat. It is shaped to the contour of the side of the boat to a thickness (approximately 2 in.) that allows the pivot arm to pass through a slot in the top side of the kayak cockpit (the coaming of a sit-in kayak) as seen in the photo of the control box and handle in Chapter 2. A belt sander works very well to contour the block. The steering mechanism and the contoured spacer are then attached to the boat with two, 3-inch long ¼ inch stainless steel screws and lock-nuts. These bolts go through the spacer and the pivot's crossbar as seen in the picture of steerer in the second photo of Chapter 2 and in the diagram above.

Installing the steering in a sit-in kayak is much easier than in a sit-upon model due to better accessibility. An opening needs to be made to reach inside the sit-upon kayak using a 5 or 6 inch deck plate. Pick a spot where both the steering spacer can be installed and where the pulley location (for the steering rope) and the rope eyes can also be reached.

The only remaining part of the steering to be built is the handle which is attached to the steering arm. Like the spacer, it is made out of a clear section of pressure treated 2x4. The cross-section is octagonal about 1 ¼ inch in diameter. First cut a 7-inch-long piece of 2x4, then rip it to 1 ¼ inches square. Then, by adjusting the fence and the angle of the blade on the bench saw, a professional looking octagon can readily be made. A ¼ inch slot is cut in one end of the handle into which the steering arm fits. The handle is locked in place with two 1 ½ inch # 6 screws with lock-nuts.

Steering Bar

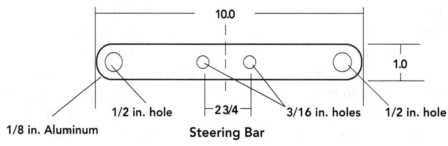

Steering Bar

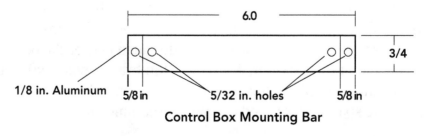

Control Box Mounting Bar

The preceding diagram and the following photo show the size of the steering bar and the location of the holes. The length should be about 10 inches to provide the optimum turning radius for the kayak. The bar is part of the 48 inch 1 ½ wide bar that was purchased for the steering mechanism. It was ripped down to one inch, but it can also be used at its full width of 1 ½ inches.

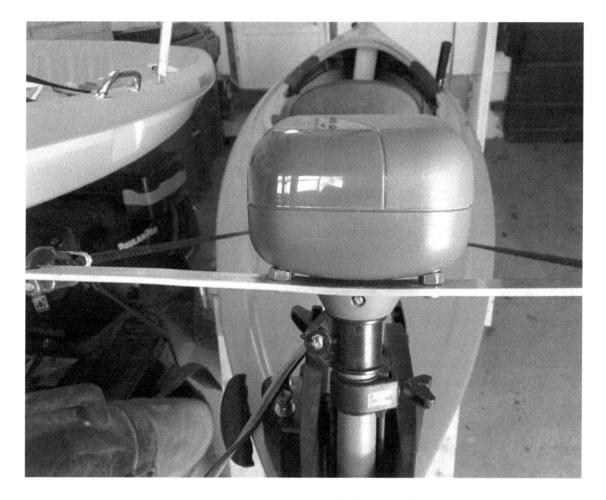

Steering Bar Installed on the N-V Trolling Motor

To attach it to the motor, the back two cover screws of the motor are replaced with screws that are ½ inch longer and spacers (a ¼ in. stainless steel nut works well) are added between the bar and the cover. The steering lines snap in the ½ in. holes at each end of the bar. See the photo above.

Mounting Bars for the Control Box

The 6 ¼ inch bars used to mount the control box are shown in the bottom part of the diagram above. They are sized for a 5 in. wide box which provides four ⅝ in. tabs extending past the bottom of the box. Two bars are needed, they are ¾ in. wide. This is another section cut from the 1 ½ aluminum bar used for steering mechanism. It was ripped in half to ¾ in.. The holes are drilled and countersunk to accommodate oval head, ¾ in. #6 screws in the center of the tabs, 5/16 in. from the ends. They attach to the 3 x 6 ¼ in. pressure treated board mounted on the sit-upon boat or on the pod of a sit-in kayak. Four flat head screws are used to attach the bars to the box one inch from the back end of the control box and ¼ in. from each edge of the box. The photos of the 2 control boxes at the beginning of Chapter 4 show the tabs.

The Mini-Transom

The likeness of a boat transom is needed to mount the electric motor on the kayak. There is just enough room for a custom built mini-transom 7 to 10 inches wide (depending on the width of the kayak) to accommodate the motor clamps which require a minimum of 6 inches in width. A good height for the transom is 3 ½ to 4 inches and a good thickness is 2 ½ inches. A block of wood of that size can be assembled from our pressure treated 2 x 4 in the following manner. Two layers of 2 x 4 and one layer of 1 x 4 are glued together to obtain the necessary height. The ends of the block are then trimmed to match the angled lines of the side of the kayak. The thickness of 2 ½ inches is cut on the bench saw. Then, the bottom of the block is contoured to match the top of the kayak as shown in the last photo of Chapter 2.

To attach the transom to the boat, I used three ¼ inch stainless steel threaded rods 2 inches longer than the height of the transom. Flat washers are needed inside the boat under the transom. Two ¼ in. rods perform double duty by bringing the motor power from the inside the boat to the surface of the transom. On the port side we have two threaded rods for the 2 wires whereas on the starboard side only one is needed. Keep the rods which power the motor separated from each other and away from the mounting bracket of the motor. Even without a battery pack inserted in the control box, it is good practice to avoid shorting the cable connections from the control box when attaching the motor.

In the sit-in kayak, the threaded rods can be pulled through from the top of the kayak using heavy sewing thread but in the sit-upon kayak, another opening must be provided to reach the bottom nuts and the power wires. A 5- or 6-inch deck inspection plate is needed here as well as for the installation of the steering and the steering lines. The least expensive inspection plates costs about $10 at West Marine or Amazon.

✎ *Major components used in this chapter are listed below.*

- Pressure treated lumber 2x4 8 feet long. Home Depot - $8

- Stainless steel threaded rod, ¼ in. diameter, 2 feet long. Home Depot - $6

- ¼ inch rope for the steering, one pulley and 6 eye bolts to guide the rope. 1 ½ in, 3/8 stainless steel bolt and lock nut. Boat supply store

- Aluminum bar 1/8 inch thick, 1 ½ in. wide 48 inches long. To be used for the steering arm, the motor control arm (ripped down to a width of 1 inch) and for the attachment of the control box to the 6 ¼ x 3 ½ inch mounting plate (ripped in half). Home Depot or Amazon - $13

- Miscellaneous stainless steel nuts, bolts and screws. Please see the text as the size sometimes depends on the kayak type.

CHAPTER 8.

In-the-Water Testing

Operation and Summary
of Performance Results

With two kayaks and two motors, we have four configurations to compare four different sets of performance results. The most important information to be gathered is the efficiency of each configuration. It will be determined by the run time that is produced by a given amount of power in watts. Equally important is the maximum speed that is achieved at each speed setting.

The following four charts show this information. For each of the six forward speed settings for each configuration, the following data is measured: the battery voltage and current (which determines the power in watts when multiplied together) and the speed. From separate tests, (the barrel tests where the motors are run in a barrel full of water) the running time is determined for each speed setting. The four configurations are as follows:

> A. *The sit-in (red) kayak with the Minn-Kota (M-K) trolling motor*
>
> B. *The sit-in kayak with the Newport Vessels (N-V) trolling motor*
>
> C. *The sit upon (coral) kayak with the Minn-Kota motor*
>
> D. *The sit upon kayak with the Newport Vessels motor*

The voltage and current readings were taken from the meters on the control box. The control box used for these tests had both a voltmeter and an ammeter. The power in watts was calculated. The speed was obtained using the satellite app "MPH" on my iPhone. The iPhone rested in a cup holder in the boat for all the tests and worked very well.

Test Results for Configuration A (Red Kayak with M-K)

Speed Setting	Motor Voltage (Volts)	Motor Current (Amps)	Power to Motor (Watts)	Boat Speed (MPH)	Running Time* (Minutes)
1	19	2.0	38	1.5	115
2	18.5	3.5	65	2.3	78
3	18	5.0	90	2.6	58
4	18	7.0	126	2.9	39
5	17.5	10.0	175	3.2	27
6	16	12.0	192	3.5	25

Test Results for Configuration B (Red Kayak with N-V)

Speed Setting	Motor Voltage (Volts)	Motor Current (Amps)	Power to Motor (Watts)	Boat Speed (MPH)	Running Time* (Minutes)
1	19	2.0	38	1.8	118
2	18.5	3.2	60	2.2	80
3	18	5.0	90	2.7	55
4	17.5	8.0	140	3.1	34
5	17.5	10.0	175	3.5	27
6	17	12.0	204	3.9	23

Test Results for Configuration C (Coral Kayak with M-K)

Speed Setting	Motor Voltage (Volts)	Motor Current (Amps)	Power to Motor (Watts)	Boat Speed (MPH)	Running Time* (Minutes)
1	19.5	1.7	34	1.7	120
2	19	3.8	72	2.3	77
3	18	5.2	94	2.5	56
4	17	7.0	119	2.8	42
5	16	9.0	144	2.9	34
6	16.5	11.0	181	3.1	27

Test Results for Configuration D (Coral Kayak with N-V)

Speed Setting	Motor Voltage (Volts)	Motor Current (Amps)	Power to Motor (Watts)	Boat Speed (MPH)	Running Time* (Minutes)
1	19	1.8	34	1.8	123
2	18	3.5	63	2.3	77
3	17.5	5.0	88	2.6	57
4	17	7.0	119	2.9	41
5	16	9.0	144	3.2	32
6	15	11.0	165	3.3	29

*The running time is determined by using the barrel tests results described below.

Barrel Tests

To determine the run time, namely, the amount of time that the motor runs continuously with one fully charged 4Ah or 6Ah battery, is a time-consuming task. Each motor must be run at 6 different speed settings for a total of nearly 7 hours when using a 6Ah battery. During this time the weather conditions are likely to change on the water. More accurate results can be obtained by running the motors under controlled conditions in a barrel full of water. There is more turbulence in the barrel but as long as the power drain from the battery is approximately the same as it is when the motor powers a boat, the results are valid. The run time numbers were obtained from the barrel tests. The curves and the two charts below show the run times for each motor.

Note that position 1a was added to the 6 forward positions for the barrel tests. It was added to produce a lower speed with less battery drain which is more compatible with position 1 when the motors are used in the open water. The tests showed that in the open water, the battery drain is considerably less than in the turbulent waters of the barrel for the same speed position. To run the barrel tests for position 1a, we simply reduced the PWM number of position 1 from 60 to 55.

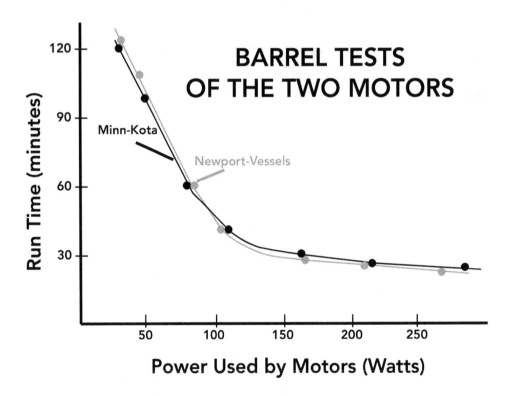

Barrel Tests Results

The following two charts show the run time of each motor for the 7 forward positions (including position 1a). This data, plotted on the curves shown above, is used to determine the run time of each position in the water test results. Batteries rated at 4 Ah were used for these tests. If 6 Ah batteries had been used the results would have been 50% greater namely 3 hours of running time for position 1a. Note that the batteries have identical characteristics using either motor.

Minn-Kota Trolling Motor

Speed Setting	Motor Voltage (Volts)	Motor Current (Amps)	Power to Motor (Watts)	Battery Discharge Time (Minutes)
1a	18.0	1.8	33	120
1	18.5	2.6	48	102
2	18.5	4.6	85	61
3	18.2	6.6	120	41
4	17.5	9.4	165	32
5	17.5	12.3	216	23
6	17.0	17.0	294	20

Newport-Vessels Trolling Motor

Speed Setting	Motor Voltage (Volts)	Motor Current (Amps)	Power to Motor (Watts)	Battery Discharge Time (Minutes)
1a	18.0	1.8	33	123
1	18.5	2.5	46	114
2	18.5	4.2	78	61
3	17.5	6.4	112	41
4	17.5	9.4	164	29
5	17.0	12.4	211	23
6	17.0	16.0	268	19

Average Speed vs. Running Time

Now that we have associated run times with boat speed (which determines the range) we can average the values for the four configurations and present the following chart with the results. For each speed setting, we have the average boat speed and the average run time generated by a fully charged 4 Ah battery.

Speed Setting	Average Boat Speed (MPH)	Average Run Time (Minutes)
1	1.7	119
2	2.3	78
3	2.6	57
4	2.9	39
5	3.2	30
6	3.5	26

Summary and Conclusions

Powering a 10-foot boat at a trolling speed of 1.7 MPH for two hours or at hull speed of 3.5 MPH for nearly one half hour with one small lithium-ion 4 Ah battery pack is a remarkable feat. I was the boat operator and I weigh 185 pounds so, with a lighter operator the results would have been even better. An average kayak paddler's speed is 2.8 MPH.

The main reason for the excellent low speed results is the use of the Pulse Width Modulation technique. It reduces the speed efficiently compared to the resistors which waste energy by converting the excess power into heat. A bonus from the PWM technique is that the software simultaneously converts the 18-volt battery pack voltage to the 12 volts required by the trolling motors.

To implement the PWM motor drive, I built four control boxes into which to plug the battery packs. The other two electronic components in the control box are the Arduino Uno which produces the PWM signal under the control of a very simple software sketch and a 20-amp Cytron driver designed to run robot motors. The Cytron requires just 3 signal wires from the Arduino Uno. Due to their simplicity, the control boxes operated flawlessly during the design, construction and testing phases of the project.

The Ryobi battery packs also worked flawlessly. I bought two 4 Ah units on sale for $100. 6 Ah units are also readily available but they cost over $100 each. Be sure to buy the genuine Ryobi product. Numerous "knock-offs" are on the market that do not meet the Ah specifications claimed. One product reviewer pointed out that these knock-offs weigh far less than the genuine product.

Unlike the lead acid batteries, the lithium ion batteries are much smaller and lighter and meet their Ah rating at all discharge rates from 2 amps to 12 amps. The built-in indicator lights are useful to monitor the charge level of the batteries. The chargers also worked very well. You can expect a full charge in about two hours.

This design emphasizes the quick setup of the propulsion system. Only two wires have to be attached to the motor. The motor attaches to the mini-transom with two large wing nuts. The steering cables snap onto the steering bar easily by tilting the motor forward without further need for adjustments when it is tilted back.

The barrel tests provided a repeatable method for building run time curves for the six speed settings. The continuous operation of an entire battery charge over several hours bodes well for the reliability of the electronics and the Cytron motor drive module which was never more than warm during the many test runs.

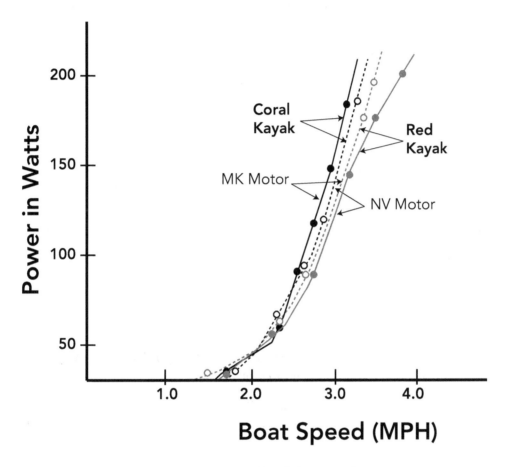

In-the-Water Test Results of the 4 Configurations

The in-the water tests results shown above in curve form for the 4 configurations indicate that the speed for the 6 speed positions respond correctly to the PWM number selected for each position. Position number 3 produces an average speed of 2.6 MPH which is close to the average paddler's speed of 2.8 MPH. At this speed, producing about ⅛ of a horsepower, a 4 Ah battery pack will drive the 10-foot kayak for 57 minutes and a distance of about 2.5 miles. The in-the-water tests also show that the conventional displacement hull of the sit-in kayak is a little faster than the flatter hull of the sit-upon kayak. The red (sit-in) kayak has a water line of 9 feet which produces a hull speed of 3 x 1.34 equal to 4 Knots or 4.6 MPH. (Square root of the water line times 1.34 in knots). We are close to reaching this speed with the N-V motor using about 200 watts of power (an input power of about ¼ of a horsepower). Overall, although the results are very close,

I would say that the red kayak with the Newport Vessels motor wins first place. The results shown on the curves above, indicate hardly any difference between any of the combinations of boats and motors between the speeds of 1.5 and 2.5 MPH when the motors consume 40 to 80 watts of power. But at 3.5 MPH (using 175 watts) both boats show that the N-V motor moves them a little more efficiently (.25 MPH faster with the same amount of power).

What Could Be Improved?

If I were to do it again, I would give more thought to making the control box waterproof. I had no water problem during the water tests, but I believe this could be a problem over the long term. Another area where the design could be improved is in the steering. It could be modified in order to make sharper turns. Increasing the distance from the pivot point to the point where the lines are attached and crisscrossing the lines where they attach to a shorter steering bar would help.

Admittedly, the panel of the control box where the battery pack plugs in, is challenging to build because the dimensions need to be exact so that the battery snaps in place and the contacts make reliably.

3D Printing of the Control Box

I built the panel shown on the following page on a 3D printer that I had just received for my birthday. I am no expert on 3D printing but with a lot of help from my grandson, Dan, we created a perfect panel.

The steps involved in 3D printing are as follows. First, the part must be modeled in 3D. While I was getting acquainted with 3D printing, I watched a dozen of hundreds of YouTube videos on the subject. For the modeling, I settled for a free program called Tinkercad. It seemed simple enough: a dozen shapes such as cubes, spheres and cylinders are available to be transformed to the dimensions of the part that you are modeling. These parts are then grouped together to make, in my case, a panel ½ inch thick of an odd shape with recessed areas and a hole in the middle.

The modeling program is then used to create an STL file representing the model. The STL file is then read by a "slicer" program (such as Cura) which tells the 3D printer how to print each layer of the part to be duplicated.

After many days of hard work I created an STL file and decided to follow the advice of the Tinkercad folks and let the experts at Shapeways (a partner company) build my model for me.

It cost me $123 and seeing that I had made a couple of mistakes in the modeling, it looked very good but was not useable.

On the other hand, grandson Dan who has a great deal of experience modeling and owns a powerful CAD program created a perfect STL file in less than an hour. Following the directions from my new 3D printer, we quickly had the machine humming. On the first try, the printer ran out of filament after 3 ½ hours (I should have been more vigilant!) and aborted, thereby ruining the print. But on the second try, after 6 hours and 53 minutes the panel (pictured right) was printed flawlessly.

3D Panel for the Control Box

I have made this STL file available, free of charge, on my web site:

myelectricboats.com/electrickayakSTL.zip.

3D printers are now available at many schools and libraries. If you provide the filament, plan ahead. The most common is PLA plastic, size 175, white. Make sure that the filament type you select can be glued to the other panels making up the control box.

The perfect solution to building the control box is to print the entire box in 3D. Dan modeled the box and created another STL file. I printed it according to the default settings of my new printer. It took 17 hours and 37 minutes and looks very good. Unfortunately, it is much too flimsy for this application.

By increasing the wall thickness to 8 mm and the density to 50%, the sturdiness of the box improved greatly but the printing time increased to 37 hours. Even with the 50% density, the walls were still too soft to use nuts and bolts for the contacts without fear of them loosening up.

Another problem was discovered when printing an entire box. As opposed to printing only a panel where the inside of the panel can be attached to the platen of the printer, when printing an entire box, the printing starts with the outside of the panel. This means that supports have to be provided for the recessed middle section of the panel. These supports are very difficult to remove cleanly after the printing is completed.

For these reasons, I recommend building the entire box of plywood or, if 3D is desired, printing only the battery panel at 100% density.

CHAPTER 9.

An Off-the-Shelf Analog Controller

Motor controllers which perform the same functions as the combination of a digital Arduino microprocessor and a Cytron driver are available off-the-shelf. Because there are so many different types of motors, there is also a great variety of motor controllers. We need to control a 12 volt, permanent magnet DC motor in the ¼ HP range. But, since our battery supply generates 18 volts DC, the controller must provide Pulse Width Modulation signals which can effectively reduce the source voltage as well as the speed of the motor.

The 20 amp Cytron motor drivers that I used for the barrel tests and the water tests were of excellent quality, they ran for hours without any problems so I ordered two Cytron MD-10-POT units (only $10 each but the shipping from Malaysia is $20). Rated at 10 amps and 30 volts, they are on the low side for this application. I also ordered an Harmnee motor controller from Amazon ($20). It is rated at 55 volts with a continuous current of 40 Amps which is more than adequate for the trolling motors. The quality, however, is only average.

Going to an analog device might appear to be going backwards in time. Certainly, digital technology is where all the action has been in the last decades. Nevertheless, some analog applications, such as the speedometer in cars, are still preferred. The motor controllers that I ordered use a proven timing device to generate the PWM pulses, the old 2N555 timer chip, which has been very reliable over the years.

For the operator, the main difference between a digital and the analog controllers is in the control device. The digital controller had distinct steps for the 6 forward and 3 reverse speeds whereas the analog controller has a continuous, volume control like knob for the speed control and a forward/off/reverse switch for the direction. Several of these controllers are available for about $20 at Amazon and ebay. This is my first experience with them so I can't recommend any of them.

For best results, purchase one that fits well in the enclosure that you will be building (they do heat up) and has long enough wires from the switch and the potentiometer so that they don't have to be cut and extended. I had no problem installing the Cytron in the control box that I built for the digital motor controllers as can be seen in the following photo.

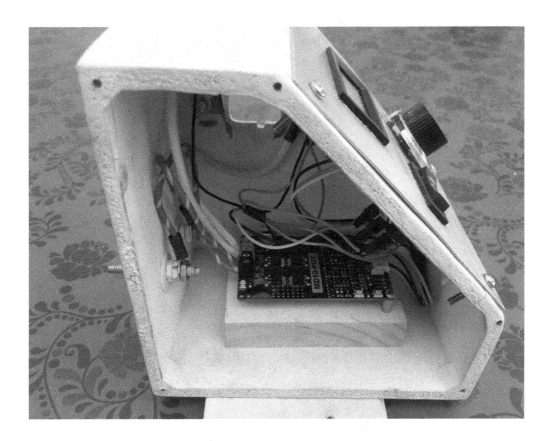

Cytron Analog PWM Motor Controller

Performance of the Analog Controllers

Before installing the new control boxes in the kayak for water testing, I connected each controller to a trolling motor for barrel testing. The results were excellent and predictable. They both troll nicely at 3 amps or less. I pushed the Cytron to 15 amps for a few minutes without ill effects and the Harmnee up to 25 amps.

The tests showed that it was easy to exceed the rating of the controllers because we are using a supply voltage of 18 volts. To avoid burning them out, it is necessary to implement stops that control the range of operation. The photo below shows a simple way to provide this safety feature. The speed control knob is replaced with one that works in conjunction with an indicator. A cut-out is then ground out for the desired range and the indicator provides a stop at each end of the safe operating range as shown in the photo below.

In the photo above, notice that there is plenty of space for the components and that the box could be narrower. The Harmnee controller (shown in the next chapter) is somewhat larger but could still fit in a slightly narrower box.

The photos in this chapter are of the 3D control box with 8 mm walls and a 50% density. This is the box that took 37 hours to print.

Knob with Cut-out to Ensure a Safe Operating Range

Water Tests

It was a windy day when my granddaughter and I did the water tests on Buckmaster Pond. I did not want to push our luck with the low power Cytron controller so we concentrated on the Harmnee which operates well within its current rating.

The test results are shown in the table below:

Off-the-Shelf Analog Controller Water Test Results

Motor Voltage (Volts)	Motor Current (Amps)	Power to Motor (Watts)	Boat Speed (MPH)
20.1	1.0	20	1.4
19.8	2.2	42	2.2
19.2	3.6	68	2.5
18.2	6.0	111	3.0
17.8	9.8	176	3.4

These results are very close to the results obtained with the Arduino digital controller as reported in Chapter 8. The analog controller is probably slightly less efficient. I noticed that it was much quieter than the digital controller. This is due to the fact that the frequency of the wave shapes (as we showed in the oscilloscope images) is much greater than the 490 Hertz that the Arduino Uno generates. When the frequency increases, there is a slight decrease in efficiency.

In conclusion, I can state that the off-the shelf analog motor controllers which use PWM do an effective job of operating the trolling motors with a minimum of work for the installation of the two controls: the speed control and the forward/ off/reverse switch. They operate more quietly than the digital controller and there is no noticeable difference in the performance.

CHAPTER 10.

Better Performance with Larger Batteries

Sizing Larger Batteries

The battery packs that we have been using are rated 4Ah and 6 Ah at 18 volts. They are capable of storing 72 and 108 watt hours of energy respectively. There are numerous battery sizes and shapes available. In the lead-acid category, a good choice for kayak, is the 12 volt batteries used in riding lawn mowers, they measure 5 x 8 inches and are 8 in. high including the height of the terminals. The best of these are rated at 35 Ah, they weigh about 17 pounds and cost $50 to $100. The capacity of these batteries is 35Ah @ 12 volts or 420 watt hours which is almost 4 times the rating of our 6Ah battery pack.

In the Lithium Ion category of 12 volt batteries, there are also many sizes. An appropriate one for a kayak would be in the 30 Ah range (360 watt-hours). I found one online that was only 3 inches wide. It weighed less than 7 pounds and cost $170. A battery such as this would provide 3.3 times the range of the 6 Ah battery pack.

So, for fishermen who want to get to the fishing hole in a hurry and still have plenty of trolling hours available, this is a good way to do it. Since we are using a 12 volt battery rather than an 18 volt battery pack, PWM is not a requirement for the controller. There are other techniques to control the speed of DC motors. The most important requirements are: high efficiency and the size of the device. I have not researched the subject to the extent that I can make recommendations, but I can say that the Harmnee controller that I bought from Amazon worked as expected and without any problems.

The 12 Volt Control Box

I had a 12 volt workshop battery, bottom of the line, Everstart riding mower battery, 230 cranking amps from Walmart ($25) which I used for these extended performance tests. A good deep cycle battery would be better and last longer but the performance results would be the same.

In the following photo, the box that I built for it is shown installed in the red kayak. The base is ½ in. plywood and the rest of the box is made of ¼ in plywood with the corners reinforced with ¼ x ¼ in. wood strips.

The motor controller and the controls are mounted in the cover so that the overall size is 6 x 9 inches, 9 in. high; it weighs 18.6 pounds. If the front of the cover were at a 45 degree angle to mount the controls without interfering with the battery terminals, it would be an improvement over this design.

The photo on the right with the open box (taken during the barrel tests) shows the Harmnee controller.

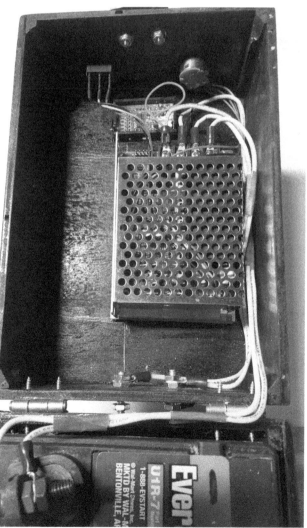

12 Volt Battery Box Showing the Harmnee Off-the-Shelf Controller

12 Volt Battery Box Installed in the Red Kayak

Water Tests

After doing a barrel test with this configuration to prove that everything worked as planned, granddaughter Hayley and I took the kayak to Buckmaster pond for the water tests and a short video.

The results of the water tests are shown in the table below.

12 Volt Battery with Harmnee Controller Water Test Results

Motor Voltage (Volts)	Motor Current (Amps)	Power to Motor (Watts)	Boat Speed (MPH)
12.3	1.6	20	1.4
12.0	5.4	65	2.5
11.6	11.6	135	3.3
11.1	19.2	213	3.7

There was no wind during these tests and we were careful to avoid the weeds. The results were slightly better than the ones obtained with the digital controller. My granddaughter who weighs much less than I do was the pilot which may account for these improved results. There was no noticeable difference with the performance results obtained with the 18 volt battery pack used for Chapter 10.

My conclusion is that it is a very good design for better performance. It can be implemented using a 12 volt, light and narrow 30 Ah Lithium battery with an off-the-shelf PMW motor controller. A box for the battery, the controller and the motor controls similar to the ones described in this book can be built as a single unit with a handle for easy lifting from the kayak. It would fit comfortably in front of the pilot.

Links to demo videos of the electric kayak in the water are available on my website, myelectricboats.com. Various clips include propulsion, steering, shifting and maneuvering in reverse, battery pack removal, as well as views of the motor/steering components while underway.

CHAPTER 11.

Transport and Launch

Part of the enjoyment of spending a day on the water in a boat is being able to deal quickly with getting the boat ready for the water. Few people have a dock in their back yard that allows them to jump in the boat anytime they feel like it. The next best thing is to have a boat on a dock or on a mooring within a short drive from home. For most of us, the reality of the expenses involved forces us to trailer or carry the boat to a beach or launching area in a car or truck.

What does not work for me is to trailer a sailboat to a launching site and then spend a long time getting the boat ready for launching. I have seen people arrive at the launch area and, even with two people doing the work, spend an hour stepping the mast, adjusting the stays, unfurling and attaching the sails, installing the outboard and launching the boat. Then, they still have to find a docking area and a place to park the car and trailer.

I have also tried the inflatable route. Even with the best pumps, hand or electric, it takes a lot of effort to get the boat ready and when you are done, you end up with a craft that's not all that seaworthy or enjoyable.

So, after trying numerous boating options, I am happy to settle for a kayak that I can handle myself and get ready for the water in a few minutes. The 10-foot kayaks discussed above weigh only 40 pounds, they are easy to load in car or truck or carried on top of the roof. I have carried canoes and aluminum boats that way but I have never been comfortable driving at highway speed with something large on my roof. The alternative is to carry the boat in the back of an SUV and leave the rear hatch partially open. This causes two problems: the interior gets dirty and the open hatch creates a lot of wind noise at highway speed. The dirt problem can only be alleviated by letting the kayak dry out for a half hour and do a good job brushing or wiping the sand off with a towel.

But the noise can be greatly reduced with a baffle. The noise level is then reduced to more like having a back window open rather than an open hatch. The picture below shows the baffle that I built for noise and wind reduction and to keep the kayak in place. The baffle is made from a 2 x 4-foot piece of ⅜ plywood.
To improve the seal between the car and the baffle, the top eight-inch section is made of 2-inch-thick foam which adapts to the contours of the back window. Home Depot sells 22 X 22-inch squares of 2-inch foam for about $5.

In my car, the baffle can be attached with two small barrel bolt latches that slide into existing holes normally used for a shelf.

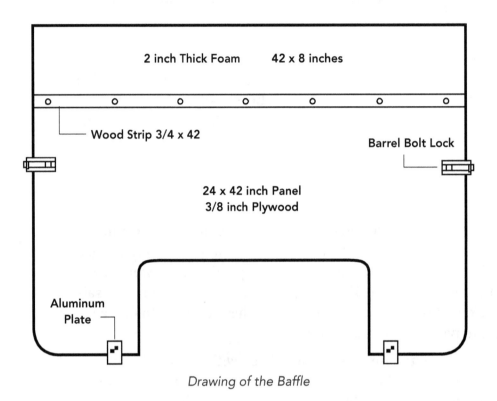

Drawing of the Baffle

The two pieces of foam approximately 8 by 21 inches each, are attached to the panel in the following manner. A slot 1 ½ in. deep is cut with an electric knife or an electric jig saw in the middle of the foam so that the foam can slide over the top part of the plywood. Eight 3/16 in. holes are drilled through the foam and the ¾ in. wood strips front and back and through the top of the plywood panel. Holes through the foam have to be punched out. I used #8 screws with lock-nuts 2 in. long (2 ½ in. long screws would have made the assembly easier). The bottom of the panel is held in place with two ⅛ in. aluminum plates measuring 1 ½ x 2 ½ inches. They fit between the carpet and the trim of the car as seen in the image on the following page.

The kayak sticks out about 18 inches beyond the hatch, which comes down to rest on the boat. A cushion between the hatch and the boat avoids scratches and rattles. Two strong shock chords are needed to keep the hatch down. In my car, an alarm sounds when driving with the hatch open. That problem is easily remedied. The alarm can be tricked into thinking that the hatch is locked down by tripping the lock mechanism into the closed position with a flat blade screwdriver. Be sure to unlatch it (using the clicker) before closing the hatch.

The Baffle in Place

Launching the Kayak

Launching the kayak is easy. Drag it to the water and secure the trolling motor to the transom with the large knobs. The power to the motor is obtained from the 2 terminals that protrude from the transom. The motor is then tipped forward and the end of the two steering lines are snapped onto the steering bar. Return the motor to the locked down position to tension the lines. A battery pack is then snapped in place and you are very quickly on your way to an exciting, fun time on the water.

Enjoy your project and visit my website, myelectricboats.com. I hope you share pics and stories of your kayak conversion!

Glossary of Terms

Ampere: a measure of the current flowing in an electrical wire

Binary Code: a numbering system that uses only 1's and 0's. Each 1 or 0 is called a "Bit." A series of 8 bits is called a "Byte"

Bus: a conductor wire to which several others are connected to form a ground bus or a power bus

C and C++: programming languages

Countersinking: drilling an angular hole which will accept a flat headed screw

Distributor: a rotating electrical device capable of connecting an input to various outputs Integer: a whole number, not a fraction

Grommet: a rubber washer which is inserter in a hole in order to insulate the wires that run through the hole from the metallic edges of the chassis

Integrated Circuit (also known as an "IC" or a "Chip"): an electronic device made up of a large number of transistors capable of performing computer functions.

Inverter: a device used to reverse the polarity of a signal. *For example, if the input signal to the inverter goes from ground to +5 volts, its output will go from + 5 volts to ground*

Jig: a device used to help machine a part precisely

Jumper: a wire with clips on each end to facilitate making a quick electrical connection

Lands: the wire equivalents embedded in the printed circuit boards

Lock-nut: an additional locking nut which prevents the original nut from loosening

Logic Board or P.C. Board: an electronic circuit board containing integrated circuits

Microprocessor: a computer processor in a single integrated circuit chip

Millivolt: one thousandth of a volt

MOSFET: electronic switching device used to drive motors or turn solenoids on and off

Nomenclature: a system of names or terms Ohm: a unit of resistance in an electrical circuit Ohmmeter: a meter to measure resistance, usually part of a multi-meter that can also measure voltage and current

Ohm's Law: Volts = amps x resistance (in ohms)

Ports, Output Ports: the computer connections which provide an output signals from the microprocessor. On this project, the signals are used to control the motor speed

Rosin Core Solder: the type of solder wire used to solder electronic parts and wire together having rosin flux built in its core

Rotary Switch: a switch that makes contact at various points as the dial is turned

Sketch: a program for an Arduino computer; the list of instructions that activate the microprocessor

Shield: a PC board designed specifically to be plugged in an Arduino computer to perform a given function such as driving motors

Shunt: a device inserted in an electrical circuit to measure current flow

Stand-offs: a plastic spacer used to separate electronic boards

Toggle: to switch back and forth between two modes of operation

Transistor: a semi-conductor device used to switch or amplify electronic signals. It is the basic building block of integrated circuits.

Volt: one volt is the voltage across a one-ohm resistor when one ampere is flowing through it.

Watt: Work done by one Amp of current multiplied by one Volt

Watt-hour: the energy generated by one watt for a period of one hour

Charles A. Mathys

Charlie Mathys was born in Brussels, Belgium and was brought to the US as a child but returned to Europe for his primary education. At age 14, he returned to the US to stay. He learned to speak English while attending high school and by age 20, he had graduated from college obtaining a BS degree in electrical engineering from Northeastern University. In 1963, he acquired an MBA degree from Boston College.

After a 2-year stint in the US Army, he joined IBM at the very beginning of the computer era. He stayed at IBM for nearly 10 years before working for three small start-up companies doing computer research and development work hoping to find fame and fortune. He finally settled down at Mitre Corp, a non-profit, "think tank" corporation doing consulting work for the Electronic Systems division of the US Air Force.

After retiring from Mitre, he combined his expertise in electronics and his love of boats to design an efficient electric motor for the propulsion of recreational boats. Several years later he designed an inexpensive RV conversion from a minivan. His skill and passion for computers and music challenged him to create a programmable modern player piano using an electronic keyboard and micro-computer. The results of his experiments combined with his previous work on electric boats gave rise to his latest electric kayak project.

He resides in Massachusetts.

Other Books by Charles A. Mathys

Electric Propulsion for Boats

After years of researching, building and testing many designs, Charlie Mathys has found the answers to inexpensive "Electric Propulsion for Boats." This book starts with an overview of his successes and failures. With each success Charlie moves ahead until you have an excellent understanding of electric propulsion for boats.

This book delves deeply into the technical aspects of electric propulsion. However the information remains extremely easy to understand. The book will take you through each phase of the required work thoroughly, with detailed explanations for each step along the way. *Visit myelectricboats.com; available on Amazon.*

My Electric Boats

"My Electric Boats" is the 2nd Edition of the popular textbook "Electric Propulsion for Boats." This updated version includes a new Rhodes 19/Etek conversion, plus more performance and efficiency tests, new photos and four complete step-by-step conversion chapters.

The examples and processes can be easily modified for small or larger vessels. If you have an interest in eco-friendly propulsion for your boat, this is the perfect place to start. *Visit myelectricboats.com; available on Amazon.*

My MiniCamper Conversion

For those who love the great outdoors, there is no greater joy than to have a versatile vehicle capable of transporting bikes, boats and camping gear to where the action is. If the same vehicle can be used for picnics, tailgate parties, music festivals and replace a second car while seating 5 comfortably, you have a real winner.The minivan conversion described in this book can do just that. It exploits the room and versatility of the minivan which, quoting Consumer Guide, provides "Easily the smartest use of space and cargo." *Available on Amazon.*

A Thoroughly Modern Player Piano

Get ready to make an affordable 'modern player piano' from an Arduino computer and keyboard. This Fun and Intensive Do-it-Yourself STEM Building Project is designed to teach the basics of Music, Robots, Electronics and Computer Hardware and Software.

Player pianos were extremely popular 100 years ago, outselling both uprights and grand pianos. The pedals were pumped to generate the vacuum that depressed the keys. The brains of the piano were the paper rolls on which the music was recorded in holes punched in the paper.

With today's technology, we can duplicate the player piano's music on a keyboard using mini-robots to activate the keys. For the brains we use a small but powerful Arduino microprocessor. This book shows how to construct the player piano using a Yamaha keyboard. It also provides the wiring charts to build a control box which connects the computer to the drivers and to the mini-robots. The software (called 'sketches' in Arduino-speak) which directs the computer to select and play the notes, is provided. *Visit modernplayerpiano.com; available on Amazon.*

www.ingramcontent.com/pod-product-compliance
Lightning Source LLC
Chambersburg PA
CBHW060205060326
40690CB00018B/4268